Probability Theory

Introduction to random variables and probability distributions

Mark Smart

Tags: probability theory and examples, probability and statistics, probability an introduction, probability theory and statistics for economists, probability for beginners, probability for finance, probabilistic graphical models, probability distributions.

Table of Contents

Disclaimer

While all attempts have been made to verify the information provided in this book, the author does assume any responsibility for errors, omissions, or contrary interpretations of the subject matter contained within. The information provided in this book is for educational and entertainment purposes only. The reader is responsible for his or her own actions and the author does not accept any responsibilities for any liabilities or damages, real or perceived, resulting from the use of this information.

The trademarks that are used are without any consent, and the publication of the trademark is without permission or backing by the trademark owner. All trademarks and brands within this book are for clarifying purposes only and are the owned by the owners themselves, not affiliated with this document.

Introduction

When making decisions about events, there are always a number of variables that need to be assessed. In most cases, we are uncertain about most of these variables. This is where probability comes in as it forms the logic for uncertainty. Probability is a good way of explaining variation and in modelling phenomena that are complex. When making decisions for uncertain events, we face the risk of making decisions that are inaccurate. It is also possible for us to make overly confident decisions. However, with probability, we are able to quantify randomness and uncertainty in an effective way. It is also a good way of strengthening intuition in cases where our initial guesses match with logical reasoning and even in cases where we are not lucky.

We face randomness and uncertainty everywhere, including when doing our daily activities as well as in science disciplines like technology and engineering. This randomness and uncertainty can only be expressed more effectively by use of probability. This means probability helps us describe random and uncertain events in the world that surrounds us. Probability is applicable in various aspects of our day-to-day lives. In most cases, we are not 100% sure about something. Through probability, one can express their degree of subjective personal belief in something. This book explores probability in detail. The author has used many practical examples to help you understand probability.

Every aspect of probability has been covered in detail. The author helps you understand how to apply probability in your day-to-day activities. Enjoy reading!

Chapter 1- What is Probability Theory

Probability theory is a branch of mathematics that deals with the analysis of phenomena that occur randomly. It is impossible to tell the outcome of a random event before it occurs, but it lies within the various possible outcomes. The determination of the actual outcome is done based on chance.

Consider the following situations:

1. Predicting the weather in New York.

2. Determining the number of students who traverse a Drill Field between 8:00 and 9:00 on Tuesdays.

3. Characterizing traffic at intersection of Road A and Road B (to determine the number of cars that traverse the intersection per unit time).

In all the above situations, we can come up with an excellent prediction or approximation for all the above experiments. However, it is impossible for us to characterize them with an absolute certainty. However, it is possible for us to characterize them probabilistically by use of probability theory elements.

You have heard weather forecasters say: the chance of having rain tomorrow is 50%. They only mean that there is a 0.5 probability that it will rain tomorrow.

Chapter 2- Basic Rules for Combining Probabilities

There are a number of rules that are used for combining probabilities. Let us discuss them:

Addition Rule

This rule can further be divided into two categories depending on whether the events being combined overlap or not: for the case of mutually exclusive events, there is no overlap. If one of the events occurs, the other events will not occur. In this case, the probability of one or more than one event occurring is obtained by getting the sum of the individual probabilities of the events.

Example:

If a fair six-sided die is thrown, the probability of any face showing up is similar to the probability of any other face showing up, which is one-sixth (1/6). There is no overlap in these six probabilities. From this we can deduce the following:

P [6] = 1/6
P [4] = 1/6
So P [6 or 4] is 1/6 + 1/6 = 1/3

The addition rule corresponds to logical *or* and it gives the *sum* of the separate probabilities.

For the case of events that are not mutually exclusive, an overlap may exist between them. We can get the probability of overlap by subtracting it from the sum of the probabilities of separate events.

If the events under consideration are not mutually exclusive and an overlap may occur between the events, addition rule may be written as follows:

P [A$\cup$ B) = P [A] + P [B] − P [A $\cap$ B]

In terms of words, the above can be expressed as, the probability of A or B or even both is sum of probabilities of A and B , minus the probability of overlap between A and B. In this case, the overlap denotes the intersection between events A and B.

Multiplication Rule

When determining the number of choices, we can follow this basic idea:

Suppose there are n_1 possible results from an operation. Each of these has n_2 possible results from the second operation. This means that there will be (n_1 * n_2) possible outcomes from the two operations together.

This means that to get the number of possible outcomes from the two operations, we multiply the number of possible outcomes at every step. To get probabilities, we can get ratios of possible outcomes.

This is the simplest form of multiplication rule for probabilities:

If the events are independent, the probability of occurrence of one event is not affected by the other event. To get the probability of more than one events occurring together, you only have to multiply their individual probabilities.

Suppose A and B are two events independent of each other. The probability of A and b occurring together can be calculated as follows:

Pr [A ∩ B] = Pr [A] × Pr [B]

If the two events are dependent, it means that the occurrence of one event is determined by the occurrence of the other event. In such a case, we must use conditional probability.

The conditional probability for event B given that A occurs can be written as P [B | A]. We read this as the probability of event B given event A, or the probability of event B given event A occurs. To find the conditional probability, we must consider only the events that meet the condition, which in our case is that A occurs.

Among the events, the probability of event B occurring is given by conditional probability P [B | A]. In a reduced sample space made up of outcomes in which A occurs, the probability of the event B is P [B | A].

If the two events are not independent, the multiplication rule for both events occurring together is the product of probability of one event and conditional probability of the second event. The following formula demonstrates this:

Pr [A ∩ B] = Pr [A] × Pr [B | A] = Pr [B] × Pr [A | B]

From the above formula, we can have the following formula for calculating the conditional probability:

Pr [B | A] = Pr [A ∩ B] / Pr [A]
Or
Pr [A | B] = Pr [A ∩ B] / Pr [B]

Permutations and Combinations

With permutations and combinations, we can get quick, algebraic methods for counting. In probability, these are used for counting the number of possible results that are equally liked for a classical approach to probability, and for counting the number of different arrangements for same items to result into a multiplying factor. Every separate arrangement or part or all of the set of items is known as a *permutation*. The number of the permutations is the number of various arrangements in which we may place the items. After changing the order of the items, the arrangement will be different; hence we will also have a different permutation.

Suppose we have a set of n items that are to be arranged, and we are allowed to choose r of the items each time, and r $\leq$ n. The total permutations of n items chosen r in time is written as nPr. In permutations, both the identity and the order of the items are of importance, hence it is considered:

nPr = n! / (n-r)! = n (n-1)(n-2)....(2)(1) / (n-r)(n-r-1)...(2)(1)

Note that 0! = 1, meaning that the number of choices for n items taken n at time t is npn = n!

The process of calculating permutations is changed in case it is impossible to distinguish some items from other items. This involves calculation of number of the permutations into classes.

If there are n different items, then the total permutations taken n at time t is n!. However, in case there are some items that can't be distinguished from the rest, then the number of permutations will be reduced. If you have n1 similar items, and the rest of the items, that is, n-n1 are the same and belong to same class, the total number of permutations will be n! /n1! (n-n1!). The numerator, that is, n!, will be the number of permutations for n distinguishable items taken n at a time t.

However, n_1 of these items cannot be distinguished, hence the number of permutations is reduced by a factor of $1 / n_1!$, while other $(n - n_1)$ items cannot be distinguished from each other, hence the number of permutations is reduced by factor of $1 / (n-n_1)!$.

If in total there are n items, where n_1 of these are of same first class, n_2 are of same second class, while n_3 are of same third class, in such a way that $n_1 + n_2 + n_3 = 1$, the total permutations will be $n! / n_1! \ n_2! \ n_3!$. This can be extended to other classes.

Combinations can be seen to be the same as permutations, but the major difference is that combinations do not consider the order of items. This means that AB and BA are not similar permutations, but they are a similar combination of letters. This means that the number of permutations is always larger compared to the number of combinations, while the ratio between the two is the number of ways in which we can arrange the chosen items.

$$nCr = \ nPr / r! = n!/(n-r)!r!$$

The nCr will give the number of ways we can choose r items from n items that are indistinguishable. These ways are equally likely. We can use this with ordinary approach to probability.

Chapter 3- Probability Distributions for Discrete Variables

Probability Functions

Suppose we have a discrete random variable X whose possible values are x0, x1, x2, ... xk and their probabilities are p(x0), p(x1), p(x2) ... p(xk). The following is true for any i:

$$p(x_i) \geq 0, and \sum_{i=0}^{\kappa} p(x_i) = 1$$
,

In which k denotes the maximum possible value for i. The p(xi) is an example of a probability function, also known as *probability mass function*. The probability function for X can also be denoted as Pr [X = xi]. In most cases, the relationship between p(xi) (or Pr[X = xi]) and the xi can be denoted using an algebraic function, but in other cases, the relationship can be shown by use of a table. We can use isolated spikes to show the relationship on bar graph. The random variable is usually represented using a capital letter such as X, while the values are represented using lower-case letters such as x, xi, x0.

Cumulative Distribution Functions

In cumulative probabilities, Pr [X ≤ x], where X is the random variable while x is an upper limit can be determined by adding the individual probabilities.

As shown below:

$$Pr[X \leq x] = \sum_{x_i \leq x} p(x_i)$$

In which p(xi) denotes an individual probability function. Example, if xi can only take a zero or a positive value:

Pr [X ≤ 3] = p(0) + p(1) + p(2) + p(3)

Expectation and Variance

The expected value or mathematical expectation of a random variable refers to the arithmetic mean that can be expected to closely approximate the mean value from a long sequence of trials after following a particular probability function. The expected value refers to the mean of all the possible results for infinite number of trials. For us to be able to calculate or determine the expectation, the complete probability function must be known. The expectation for a random variable X can be expressed using E(X) or μx or μ. The two last symbols are an indication that the expected value or expectation is the mean value for the distribution of our random variable. This can be expressed using the formula given below:

$$E(X) = \mu_x = \sum_{all\ x_i} (x_i)Pr[x_i]$$

Variance of Discrete Random Variable

The variance is normally obtained as (xi − µ)². The quantity that corresponds to the mean for probability distribution is the expectation; hence the variance for the discrete random variable must meet the following:

$$\sigma_x^2 = E(x - \mu_x)^2$$
$$= \sum_i (x_i - \mu_x)^2 Pr[x_i]$$

Chapter 4- Binomial Distribution

The binomial distribution is normally used in cases where an experiment gives two possibilities, a success and a failure. It is a discrete probability distribution that expresses a set of two alternative successes (p) and the failures (q).

The following are the requirements for us to use the binomial distribution:

The determination of the output completely relies on chance.

The experiment has two possible outcomes only.

All the trials have same probability for a certain outcome in one trial. This means that the probability for a subsequent trial is not dependent on the outcome of the previous trial. Assume the constant probability for one trial is p.

The total number of trials, that is, n, must be fixed, despite the outcome of every trial.

The following probability function defines binomial distribution:

$$P(X-x) = {}^{n}C_{x}Q^{n-x}.p^{x}$$

Whereby:

- p is the probability of success.
- Q is the probability of failure, that is, 1−p.
- N is the number of trials.
- P(X−x) is the probability of achieving x successes for n trials.

Example 1:

Suppose a test with multiple choice questions is conducted. Each question has 4 choices for the answers, out of which only one of them is correct. Calculate the probability that a person who undertakes the test scores 5 questions wrong.

Solution:

From the above information, the following are the values for the various variables:

n = 20,
n - k = 5,
k = 20 - 5 = 15

The probability of giving the right answer is the probability of success, s = ¼ - we have 4 options for answer per question. The probability of failure is simply the probability of giving wrong answer, which is, 1 −s, which is, 1 − ¼ = ¾. The above can then be substituted in the formula for calculating the binomial distribution to give the following:

P (probability of getting 5 incorrect answers out of 20) =

C (20, 5) * (14) ^{15} * $(3/4)^5$
P = [(20*19*18*17*16)/ (5*4*3*2*1)] * $(1/4)^{15}$ * $(3/4)^5$
= 0.0000034

This means that we need a probability of approximately 0.0000034.

Example 2:

Suppose we are tossing a coin three times. Calculate the probability of getting THREE heads.

Solution:

We are tossing the coin for three times. H represents a Head while T represents a Tail. These are the two possible outcomes from the experiment. The following are the possible outcomes from the three trials:

HHH, HHT, HTH, HTT, THH, THT, TTH, TTT

We need an output with two heads, and these may appear in any order. These are possible outputs with two heads: "HHT", "THH" and "HTH". This means that only 3 of the total outputs contain two heads. Now, what is the probability for every output? All the outputs can be said to be equally likely. We have 8 outcomes, meaning each of them has a probability of 1/8.

The probability of getting two heads can be calculated as follows:

Number of outcomes we want x Probability of each outcome

= 3 × 1/8 = 3/8

This means that the probability of getting two heads after tossing the coin three times is 3/8.

Example 3:

Suppose we need to calculate the probability of 5 Heads in a total of 9 tosses. It will be tiresome for us to list down all the 512 outcomes. We can easily solve this problem after creating a formula.

Here is the formula that we can use:

$$\binom{n}{k} = \frac{n!}{k!(n-k)!}$$

Which can be read as "n choose k", whereby:

- n is the total number
- k is the number we need
- the "!" symbol represents factorial.

Example 4:

Calculate the probability of getting 2 Heads in three tosses.

Solution:

From the above information, we have the following:
n = 3
k = 2
n!k!(n-k)! =   3! / [2!(3-2)!]
= (3×2×1) / (2×1 × 1)
= 3

This means that there will be 3 outcomes with 2 heads.

Example 5:

The probability of getting is 5 Heads in 9 tosses.

In this case, we have the following information:

n = 9
k = 5
= 9! / 5!(9-5)!
=9×8×7×6×5×4×3×2×1 / **5×4×3×2×1 × 4×3×2×1**
= 126

From the result obtained above, a total of 126 outcomes from the experiment will have 5 heads.

We should now determine the total number of outcomes in the 9 tosses. This can be calculated as follows:

$2^9 = 512$

Since we have the above information, the probability can now be calculated as follows;

Number of outcomes we need x Probability of each outcome

$126 \times \ 1/512 = 126/512$

This means that:

$P(X=5) = 126/512 \ = 0.24609375$

After rounding off the above result, we conclude that there is a 25% probability of getting 5 heads in 9 tosses.

Example 6:

Suppose your company manufactures sports bikes. 90% of these bikes pass the final inspection, while 10% fail and should be fixed. Calculate the expected variance and mean for the next 4 inspections.

Solution:

From the above information, we can deduce the following:

- $n = 4$,
- $p = P(\text{Pass}) = 0.9$

X is a random variable denoting the number of passes from the four inspections.

Substitute in the formula for x from 0 to 4. This gives us the following:

P(k out of n) = $[n! / k!(n-k)!]p^k(1-p)^{(n-k)}$

We can then substitute the values of k from 0 to 4 as shown below:

$P(X = 0)$

We substitute k with 0:
$= 4! / 0!4! \times 0.9^0 0.1^4 = 1 \times 1 \times 0.0001 = 0.0001$
$P(X = 1)$

We substitute k with 1:
$= 4! / 1!3! \times 0.9^1 0.1^3 = 4 \times 0.9 \times 0.001 = 0.0036$
$P(X = 2)$

We substitute k with 2:
$= 4! / 2!2! \times 0.9^2 0.1^2 = 6 \times 0.81 \times 0.01 = 0.0486$
$P(X = 3)$

We substitute k with 3:
$= 4! / 3!1! \times 0.9^3 0.1^1 = 4 \times 0.729 \times 0.1 = 0.2916$
$P(X = 4)$

We substitute k with 4:

$$= 4! / 4!0! \times 0.9^4 0.1^0 = 1 \times 0.6561 \times 1 = 0.6561$$

After that, we can interpret our results. It means that for the 4 next bikes, there is a 0.01% chance of having no pass, 0.36% chance of having 1 pass, 5% chance of having 2 passes, 29% chance of having 3 passes and 66% chance that all the 4 bikes will pass the inspection.

Chapter 5- Poisson Distribution

Poisson is a discrete distribution used when particular conditions are met. It is also applicable in the binomial distribution as an approximation method. The Poisson distribution is well applicable in situations where the number of occurrences that are possible is larger compared to the average number of occurrences in a particular interval of space or time. The total number of possible occurrences is usually not known exactly. The occurrence of the outcomes must also be random, that is, by chance, and the probability of occurrence of an item should not be affected by whether the item occurred previously, meaning that the occurrences are independent.

Calculating Poisson Probabilities

The probability of getting exactly r occurrences within a fixed interval of space or time under certain conditions is given by the following formula:

$$\Pr[R = r] = \frac{(\lambda t)^r e^{-\lambda t}}{r!}$$

Where t represents the interval of space or time and is measurable in units of length, time, volume or area. λ denotes the mean rate of occurrence per unit space or time. This means that the product of λt will be dimensionless. The e forms the base for natural logarithms, with an approximate value of 2.71828.

With a Poisson distribution, it is possible for us to predict the likelihood of an event occurring given how it has occurred previously. It normally gives us the probability of a particular number of events occurring within a fixed duration of time. If you need to predict the likelihood of events occurring, and the events have a low probability, use the Poisson distribution.

Example 1:

2.5 goals are scored on average per game in the premier league. Using the Poisson distribution, calculate the probability that K goals will be scored in a game.

Solution:

We will use the following formula:

$$P(X = k) = \frac{\lambda^k e^{-\lambda}}{k!}$$

In the formula, e denotes the Euler's number. The value of λ is 2.5, that is:

$\lambda = 2.5$

$P(X=0) = 2.5^0 e^{-2.5} / 0!$
$= 0.082$

$P(X=1) = 2.5^1 e^{-2.5} / 1!$
$= 0.205$

$P(X=2) = 2.5^2 e^{-2.5} / 2!$

$= 0.257$

$P(X=3) = 2.5^3 e^{-2.5} / 3!$

$= 0.213$

$P(X=4) = 2.5^4 e^{-2.5} / 4!$

0.133

You can then plot a bar graph showing the values of X on the X axis, that is, 0, 1, 2, 3 and 4. The values of probabilities obtained above should be plotted on the Y-axis.

The formula has no upper limit for the value of k, but the probability approaches 0 rapidly with the increase in the value of k.

Example 2:

A mathematician observes the number of cars that approach a junction. He notices that an average of 1.6 cars approaches the junction every minute. Supposing is that the number of cars approaching the junction follow a Poisson distribution, calculate the probability that 3 or more cars approach the junction every minute.

Solution:

In this case, the value of λ is 1.6, that is: $\lambda = 1.6$

Our goal is then to find the following:

$P(X \geq 3)$

Which is the probability that we have 3 or more cars approaching the junction every minute? Since the value of k does not have an upper limit, we can directly calculate the probability. However, after calculating the complement, that is $P(X \leq 2)$, we can easily get the value for $P(X \geq 3)$.

So let us calculate the $P(X \leq 2)$:

$P(X = 0)$
$= 1.6^0 e^{-1.6} / 0!$
$= 0.202$

$P(X = 1)$
$= 1.6^1 e^{-1.6} / 1!$
$= 0.323$

$P(X = 2)$
$= 1.6^2 e^{-1.6} / 2!$
$= 0.258$

To get the probability $P(X \leq 2)$, we can add the probabilities of X at 0, 1 and 2. That is:

$P(X \leq 2) = P(X = 0) + P(X = 1) + P(X = 2)$

= 0.202 + 0.323 + 0.258

= 0.783

Now that we have the above, we can determine the P(X ≥ 3 as follows:

P(X ≥ 3) = 1 - P(X ≤ 2)

= 1 – 0.783

= 0.217

The above answer means that the probability of having 3 or more cars approach the junction every minute is 0.217.

Example 3:

An average of 4.5 calls is received in a customer care desk for every 5 minutes. Each customer care can handle only 1 call for the 5 minutes. If a customer calls and customer care is available is to receive the call, the caller/customer is placed on hold. Suppose that the calls follow Poisson distribution, determine the minimum number of agents that are needed to be on duty so that at most only 10% of the calls are placed on hold.

Solution:

We have an average of 4.5 calls per 5 minutes. This means:

$\lambda = 5$

Let us start our calculations:

$P(X = 0)$
$= 4.5^0 e^{-4.5} / 0!$
$= 0.011$

$P(X = 1)$
$= 4.5^1 e^{-4.5} / 1!$
$= 0.061$

$P(X = 2)$
$= 4.5^2 e^{-4.5} / 2!$
$= 0.173$

$P(X = 3)$
$= 4.5^3 e^{-4.5} / 3!$
$= 0.342$

$P(X = 4)$
$= 4.5^4 e^{-4.5} / 4!$
$= 0.532$

$P(X = 5)$
$= 4.5^5 e^{-4.5} / 5!$
$= 0.703$

$P(X = 6)$
$= 4.5^6 e^{-4.5} / 6!$
$= 0.831$

$P(X = 7)$

$$= 4.5^7 e^{-4.5} / 7!$$
$$= 0.913$$

This result can be interpreted as follows:

If we need to ensure that no more than 10% of calls are kept on hold per unit time, then 7 customer cares are requ8ired to be on duty.

Chapter 6- Normal Probability Distributions

These types of probability distributions are very popular in statistics. They result into a very normal curve. Examples of these are when you are measuring the heights or weights of individuals, or getting opinions from individuals.

A normal distribution can be said to occur naturally and it is sometimes it is known as a *bell curve*. Once students do an exam such as GRE, most of them will score an average grade of C. A small number of students will score B while others will score grade D. Another even smaller number of students will score grades A and F. This results into a normal distribution that looks like a bell, hence the name *bell curve*. A bell curve is normally symmetrical, meaning that half of the data falls to the left of the mean while half of the data falls to the right of the mean.

This is the pattern followed in majority of the groups. This explains why this type of distributions is highly applied in statistics, businesses and government bodies. The following are the properties that characterize a normal distribution:

- The mean, the median and the mode are all equal.

- It has a symmetrical curve at the center.

- Exactly half of all values are to the right of the center while the other half are to the left of the center.

• The region under the curve has a total area of 1.

In normal distributions, we are only interested in two parameters, that is, M and Σ^2.

If we have a continuous normal variable X, its probability in the interval [a, b] is simply the area under the curve and bounded by x=a and x=b, which can be calculated as follows:

$$P(\alpha < X < b) = \int_{\alpha}^{b} f(X)dx$$

The area will depend on the value for M and Σ.

After standardizing the normal curve, life becomes easier, with the mean being 0 and standard deviation 1 unit.

Suppose we have a standardized solution with the following values:

M = 0 and
Σ = 1

We will finally have the following:

$$f(X) = \frac{1}{\sqrt{2\pi}} e^{-x^2/2}$$

All observations for a normal random variable X having a mean M and variance Σ can be transformed to new set of observations for another random normal random variable z with a mean of 0 and a variance of 1 by use of the transformation given below:

$$Z = \frac{X - \mu}{\sigma}$$

EXAMPLE 1:

Let X is the IQ of Americans chosen randomly. X ~ N(100, 162). Calculate the probability that an American chosen randomly has an IQ below 90.

Solution:

We first use this formula:

$$Z = \frac{X - \mu}{\sigma}$$

Z = (90 − 100) / 16
= -0.63

This should give a probability of 0.2643.

The following is true:

P(X < 90) = P(Z < −0.63) = P(Z > 0.63) = 1 − P(Z < 0.63)

The first equality comes from transformation of X to Z, the second one from symmetry of normal distribution while the third one is from the rule for complementary events. The probability of P(Z < 0.63) is 0.7357 as from the table. Since we know this, we can deduce the following:

P(X < 90) = 1 − P(Z < 0.63) = 1 − 0.7357 = 0.2643

The above shows that we will always arrive to the same answer despite the method that we use.

Suppose we need to calculate the probability that any American picked randomly has an IQ of above 140. We can proceed as follows:

$$Z = \frac{X - \mu}{\sigma}$$

(140 − 100) / 16
= 2.50

You can check from the table and you will get the probability, that is, the probability that after picking an American randomly, they will have an IQ of above 140.

Alternatively, we can arrive to the same solution using the following method:

P(X > 140) = P(Z > 2.50) = 1 − P(Z < 2.50)

The first equality comes from transformation of X to Z, while the second one is from the rule for complementary events. From the table, the P(Z < 2.50) is 0.9938. This means:

P(X > 140) = 1 − P(Z < 2.50) = 1 − 0.9938 = 0.0062

Again, we have used two different methods but we have arrived at the same answer.

Again, we can calculate the probability that an American picked randomly has an IQ ranging between 92 and 114.

Solution:

= (92 − 100) / 16
= -0.50
= (114 − 100) / 16
= 0.88

P(92<X<114) = P(-0.50<Z<0.88)
= P(Z<0.88) − P(Z0.50)
= 0.5021

This shows that the probability than any American who is picked randomly will have an IQ ranging between 92 and 114.

Now you are able to find the probabilities above a number, below a number and between any two numbers.

At this point, you should be in a position to find any probability when you are asked.

Finding X

At this point, you know how to use standard normal curve N(0, 1) to get the probability of a normal random variable X with a mean M and a standard deviation Σ. In some cases, we may not be interested in finding the probability but the value of the variable X. It is possible that we use probability to find the value of a normal random variable X. This is what we shall be discussing in this section.

Example:

Suppose a test has been done and the teacher wants to distribute the marks with a mean of 70 and a standard deviation of 10. The teacher wants 15% of the class to be awarded an A. Determine the cutoff that the teacher should use to score grade A.

Solution:

From the table, one can easily see that the value of Z is 1.04, hence we have the following:

Z = 1.04

$$Z = \frac{X - \mu}{\sigma}$$

$X = \mu + Z\sigma$
$X = 70 + 1.04\,(10)$
$X = 80.4$

To make it easy for your understanding, whenever you need to find a normal random variable X using normal probability, follow these steps:

• Find the value of Z associated with it, that is, the normal probability.

• Using the transformation $X = \mu + Z\sigma$, find the value for random variable X.

Chapter 7- Sampling

A sample refers to a group of readings or objects that have been taken from a population for measurement or counting purposes. The properties of the entire population are inferred from the observations made from the sample. A good example is the mean of the sample, x, which is an unbiased mean of the population, μ. The variance of the sample s2 is also an unbiased estimation of the variance of the population, $\sigma2$.

A sample should be random, meaning that every element in the population must be given an equal chance to be chosen and be part of the sample. This way, there will be no bias in the sampling process. If you are investing a factor or a number of factors and it is feared that the rest of the factors may interfere, the sampling process must be done carefully to avoid bias from the interfering factors. You can minimize the impact of the interfering factors or make them insignificant by randomizing with a great care both the order in which the items are taken for sampling and the choice of sample parts that receive different treatments.

Types of Probability Sampling

In probability sampling, each member of the population has an equal chance of being selected into the sample. The amount of chance is also known. For example, if there are 50 elements in the population, the chance of each element being selected into the sample is 1 out of 50.

However, in a non-probability sampling, this is not the case. For example, an individual who lives near a researcher has a high chance of being interviewed by the researcher compared to an individual who stays far from the researcher. This means that with a probability sampling, there are high chances of coming up with a sample that is a true reflection of the entire population.

There are different types of probability sampling. Let us discuss them:

1. Simple random sampling

This is the easiest method for probability sampling. When performing simple random sampling, the researcher should include all the population members into a master list, and then the elements of the sample are chosen randomly from this list. This way, one can come up with a sample that truly represents the population.

The objective in simple random sampling is to choose n units from N items such that $_NC_N$ has equal chances of being chosen. For example:

Suppose a firm wants to interview its customers regarding the quality of service they receive from firm services over the past one year. The question is, how can a simple random sample be selected? First, the sampling frame should be organized. The company should go through its records and identify their clients for the past 12 months. A decision should be made to know the number of clients that should be included in the final sample. If there were 1000 clients for the past 12 months, you can decide to choose a sample of 100 clients. This will mean we will have a sampling fraction of f = n/N = 100/1000 = .10 or 10%.

To choose the sample, you can decide to print all the 1000 clients, create strips for each client. The strips can be added into a container then mixed thoroughly. The first 100 strips can be picked and be used in the sample. However, this can be a very tedious procedure. The quality of the sample will also be determined by how thoroughly you mix the strips and how you pick them from the container.

2. Stratified Random Sampling

This is also known as the proportional random sampling. In this sampling technique, the population elements are first grouped into a number of classifications like level of education, gender or their socioeconomic status. However, the classifications must not have overlapping subjects. Once the classes have been created, the researcher goes ahead to choose elements from the various classes and be used in the sample. This ensures that we have a sample that is a true representation of the entire population.

Stratified random sampling is well applicable where the goal of the researcher is to study a certain subgroup within a large population. Stratified random sampling gives more precise statistical measures compared to simple random sampling. It works by creating layers within the population which can give a more accurate sample, but the method can be tedious and time consuming.

3. Systematic Random Sampling

This sampling technique can be seen to be the same to an arithmetic progression where any two consecutive values give the same difference. For example, if you need to do a research on a class of 100 students, you will first choose a number that is less than 100, that is, the total population. This should form the first step in systematic random sampling. We can pick number 4 as the subject. We should then select another integer, which is the number of individuals we can have between the subjects. We can choose 6 for this. If we follow the numbers chosen above, then the students chosen for the study will be 4, 10, 16, 22, 28...With systematic random sampling, it is easy for one to create a sample without having to rely on a random number generator. However, when using this technique, the results will not be as random as when using a random number generator.

4. Cluster Random Sampling

This sampling technique is used in cases where the population is too large for use of simple random sampling technique. Suppose you need to study the lifestyle in Europe. This is a very huge population.

In this sampling technique, the first research has to establish the boundaries. In our above example, the boundaries will be the various countries in Europe. The researcher then goes ahead to choose a number of boundaries that has been identified. Each of these, that is, the various countries in Europe, should have equal chances of begin selected.

Once the countries have been chosen, the researcher may decide to use all the individuals in that country or use simple random sampling to select only a few individuals from that country.

Cluster random sampling is easy to use and offers a great convenience to the researcher. However, if the individuals involved in the research are not homogenous, then the researcher will end up getting data that is less accurate. It may also be inaccurate to make conclusions from such data.

5. Multi-stage Sampling

This probability sampling technique involves the combination of any two or more sampling techniques discussed above. A single sampling technique is not enough to get the level of randomization needed to achieve accurate results. When two or more sampling techniques are used, one gains confidence that the data they get from the research is more accurate and can be used for making sound decisions.

It is also a good way of minimizing bias in creating a sample from the entire population.

Linear Combination of Variables

Suppose we have X and Y which are two independent variables. A linear combination is made up of sum of a constant then multiplied by one of the variables then another constant multiplied by the remaining variable. This can be expressed as follows in terms of algebra:

W = aX + bY

In which W is the combined variable while *a* and *b* are the constants.

The mean for a linear combination is simple as follows:

W = aX + bY

After multiplying a variable by a constant, its variance increases by a factor of the constant squared, that is, *variance (aX) = a2 variance(X)*. This is consistent since the variable will have units of the square of your variable. After a combination of two variables, the variances must increase. A cancellation cannot happen since variability's accumulate. The value of variance must always be positive, meaning that once it is multiplied by a constant, the result must be a positive value. This means that there is sense in the combination of the two independent variables:

$$\sigma_w{}^2 = a^2\sigma_x{}^2 + b^2\sigma_\gamma{}^2$$

If we need to combine over two independent variables, the process can be done in a similar way. If the two independent variables X and Y are just added together, the constants a and b will be both equal to one, so we should add the individual variances as follows:

$$\sigma_{(X+Y)}{}^2 = \sigma_x{}^2 + \sigma_\gamma{}^2$$

If our variable W is a sum of n independent variables X, each with same probability distribution and the same variance $\sigma x2$, then we can have the following:

$$\sigma_w{}^2 = (\sigma_x{}^2)_1 + (\sigma_x{}^2)_2 + ... + (\sigma_x{}^2)_n = n\sigma_x{}^2$$

Variance for Sample Means

This helps us know how reliable the sample mean is as an estimate of population mean. Suppose we have a sample with n independent observations. Suppose every observation is multiplied by 1/n, the sum of these products will be the mean of observations.

Central Limit Theorem

If independent and random samples are taken from a population having a mean of μ and a variance of $\sigma2$, an increase in the sample size n, the distribution of the sample means will approach a normal distribution.

The mean for the sampling distribution will be μ while its variance will be σ2/n. How large is the sample size needed to be before the sample mean distribution becomes approximately a normal distribution?

This will be determined by the shape of original distribution. If the original population has a normal distribution, the means for samples of any size will have a normal distribution. The differences and sums for variables that are normally distributed will also have a normal distribution. If the distribution of the original population was not normal, the means for samples of size two or larger will get closer to normal distribution. If the samples are taken from all the distributions that are encountered practically, the sample means will have a normal distribution and have a negligible error if the size of the sample is not less than 30. Exceptions exist only in samples that have been taken from populations with distant outliers.

Chapter 8- Applications of Probability

Probability is used to measure the likelihood of a certain event occurring. The probability theory is a very important subject that can be applied in various fields. Let us discuss the various applications of probability in everyday life:

1. Risk estimation

Insurance companies are applying probability on a daily basis for risk estimation. These companies are also using probability and statistics to make marketing decisions and determining prices.

The insurance companies also rely on probability to process policy applications. Example, policyholders who smoke are at a high risk of developing serious health complications. This normally leads to increased insurance claims. The age and geographic location of the applicant can serve as a basis for predicting future claims.

2. Annuities and Life Assurance

To analyze mortality rates, the insurer assesses the residence of the applicant and the socioeconomic factors that apply to the current age and health of the policyholder. Such an analysis will help the insurer determine the rates as well as options for annuities and life insurance policies. Probability theory can help the insurer to predict the number of years that the applicant or policyholder will live.

3. Liability and Property

Companies providing liability and property insurance rely on probability for risk assessment. Research has shown that the age and gender of a driver determines the likelihood of an accident occurring. The insurer also considers the type of the insured vehicle, the geographic location of the driver and the distance driven regularly when setting the premium rates based on probability. When insuring a home, the insurer will consider factors such as the location, the heating system used in the home, the age of the property and any security measures used to secure the home. This way, the insurer is able to come up with a fair rate that the home owner will be paying on a monthly or annual basis.

4. Weather forecasters rely on probability to make predictions based on past observations. It is always hard to predict how the weather will be like in the coming hours, days, weeks and even months. Making a prediction for each of these durations comes with its own challenges. Everybody wants to be told how the weather will be like so that they may make the right decisions based on the location they want to be in during that time. However, nature doesn't operate this way, but we can rely on probability to tell people on the probability of something happening.

With probability, we can predict how the weather will be like with a high degree of accuracy. The forecasters can combine computer generated forecasters with their own experience to predict how the weather will be in the future.

5. Sport betting

Sports betting are an area where probability can be applied successfully. Although the predictions of the sport may not be published much, they are good in determining the pay-off rates.

6. Economic Forecasting

This is the process of forecasting the various factors of the economy such as inflation and GDP (Gross Domestic Product). Point forecasts do not consider uncertainties, but probability comes in to help. Probability can help a government to predict various aspects of its economy to take the appropriate action in advance.

7. Biology

Probability can be used for assessing biology trends such as the spread of diseases. This can help in predicting the probability of a disease occurring in the future and take the necessary action early in advance. This way, we can avoid the spread of diseases and even reduce the number of deaths brought about by certain diseases.

8. Energy Forecasting

Although not much has been done in energy forecasting with probability, but this is changing. The Global Energy Forecasting Competition (GEFCom) is aiming at probabilistic forecasting of wind power and electric load, electricity pipes and solar power.

9. Speech Recognition

In speech recognition, probability is used or applied in determining the most likely spoken word in past speeches. This is the word with the highest probability of being spoken. This can be done for a specific speaker, with the end result being better speech recognition for that speaker. A system that finds it hard to understand the speech from a certain individual can finally find it possible to understand the speaker.

10. Most standards for image recognition rely on unequal probabilities of their pixel densities. An example of this is jpeg.

11. Probability theory is also highly applied in reliability. Most consumer products like automobiles and consumer electronics rely on probability theory to design the products with a goal of reducing the probability of failure. This is of great help to the manufacturer when making decisions about the product's warranty.

All the above points are a clear indication that probability is very applicable in our day-to-day activities in our lives. When you are travelling to a certain destination, probability can help you know the most effective route to that destination based on a number of factors including traffic jam.

Conclusion

This marks the end of this guide. Probability theory helps in determining the likelihood of an event occurring. The probability theory is one of the branches of mathematics that deals with the analysis of phenomena that occur randomly. It is impossible to tell the outcome of a random event before it occurs, but it lies within the various possible outcomes. The determination of the actual outcome is done based on chance.

There are various probability distributions. The best one to apply is determined by the way in which your data is distributed. There are also rules that govern how probabilities should be added, multiplied. These rules also govern how permutations and combinations can be carried out on probabilities.

Probabilities can be applied in everyday life. For instance, the probability theory can be applied in weather forecasting to predict how the weather will be like in the coming hours, weeks, or even months. Insurance industries rely on probability when making most of their decisions. This helps them calculate the amount of payment that the policyholder should pay per month or on an annual basis. Governments can use probability theory to predict various factors of their economy. A good example of this is inflation.